I0055504

STEAM QUEST

Story-Based Activity Book

By AHMYA RIVERA

STEAM Quest: Story-based Activity Book by Ahmya Rivera

Published by: JilleyBean Books, Inc.

Website:

Copyright © 2022, Ahmya Rivera

All rights reserved. No portion of this book may be reproduced in any form without permission from the publisher, except as permitted by U.S. copyright law. For permissions contact: Jilleybeanbooks@gmail.com

Cover by: Garin Adi

Dedicated to all the young leaders who haven't unlocked their true potential... yet.

⚠️

Disclaimer

One reason for the creation of STEAM Quest is to provide representation of different groups through engaging activities. To partake in these engaging activities we highly recommend supervision from someone 18+ if the individual conducting the activity is under 14 years old.

Materials

All the materials needed to conduct every activity within this book.

-Scissor (x1)

-Gluestick (x1)

-Pencil (x1)

-Popsicle sticks (Jumbo and Regular)

-Crayons or Markers

-Pipe Cleaners(x5+)

-Glass cup (x1)

-Very hot water (3 cups)

-Borax (2 cups)

-Popsicle stick (x50+)

-String (5inch long)

-Liquid Glue (x1 container)

-Stapler (x1)

-Ruler (x1)

-Colored paper (x5: Green is essential)

-Plastic plate (x1)

-Fake lightbulb candle (x1)

Optional: Food coloring

Amora has a never-ending wish to create. Some even call her a "Little Mechanic". Creating new equipment from older ones was her specialty! While other students were playing, she was creating new things... and her new creation was the best thus far.

1

After months of drawing, planning, and building, it was finally done. Amora started packing her bag! She grabbed a screwdriver, scissors, wires, a magnifying glass, a globe, and more. the Little Mechanic was preparing for her first journey, and she didn't want to wait any longer.

Her newest creation, a train, would allow her to see and try new things. this was the moment Amora had been waiting for! And she finally built up the courage to make it happen, even when others said she couldn't.

So, without hesitation, she grabbed her backpack and quickly jumped into the train to start her adventure.

4

Amoras' imagination didn't stop while on a journey. In fact, a new idea had been on her mind for quite some time. She had picked up tubes, metal, and some other materials. All of these items had been collected for one purpose -- to create a travel buddy!

His name would be "Steam" the robot. Just like Amora, he is highly knowledgeable and isn't afraid to show it. Which, undoubtedly, came in handy while on their quest

It's time for you to build...

STEAM!

Materials: Scissors, a gluestick, a pencil, two regular popsicle sticks, crayons or markers, and a pipe cleaner.

Step 1: Cut out all the pieces to the right.

Step 2: Color the sections however you want.

Step 3: Fold the paper in accordance with the photo on the left.

Step 4: Use the scissors to poke through the blue solid lines.

Step 5: Make a marker to draw a ""U"" pattern on both sides of two sets of cut in half popsicle sticks using a marker.

Step 6: Stick the popsicle sticks in the holes as the legs and arms, and the pipe cleaner on the whole at the top of Steam's head.

Step 7: Glue the star onto the top of the pipe cleaner. Then you're done!

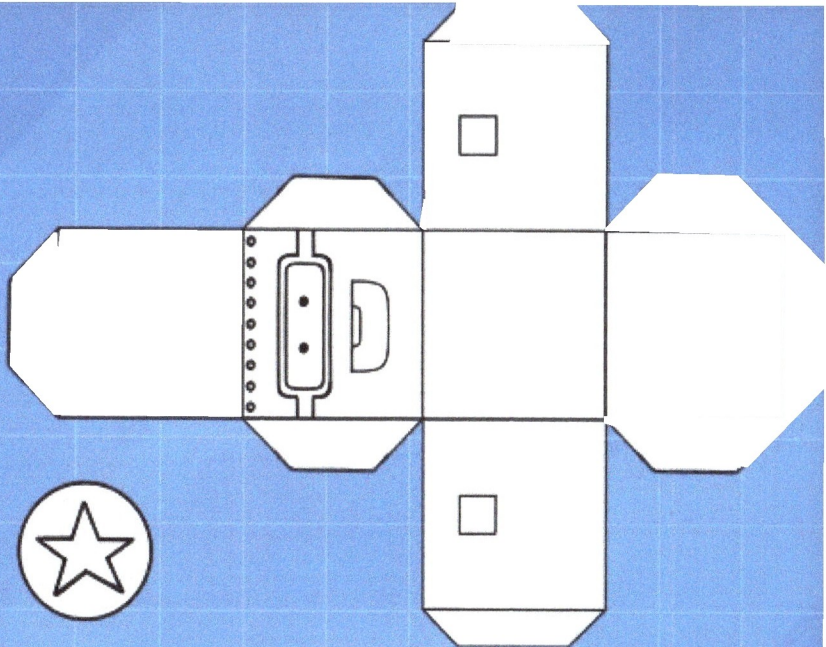

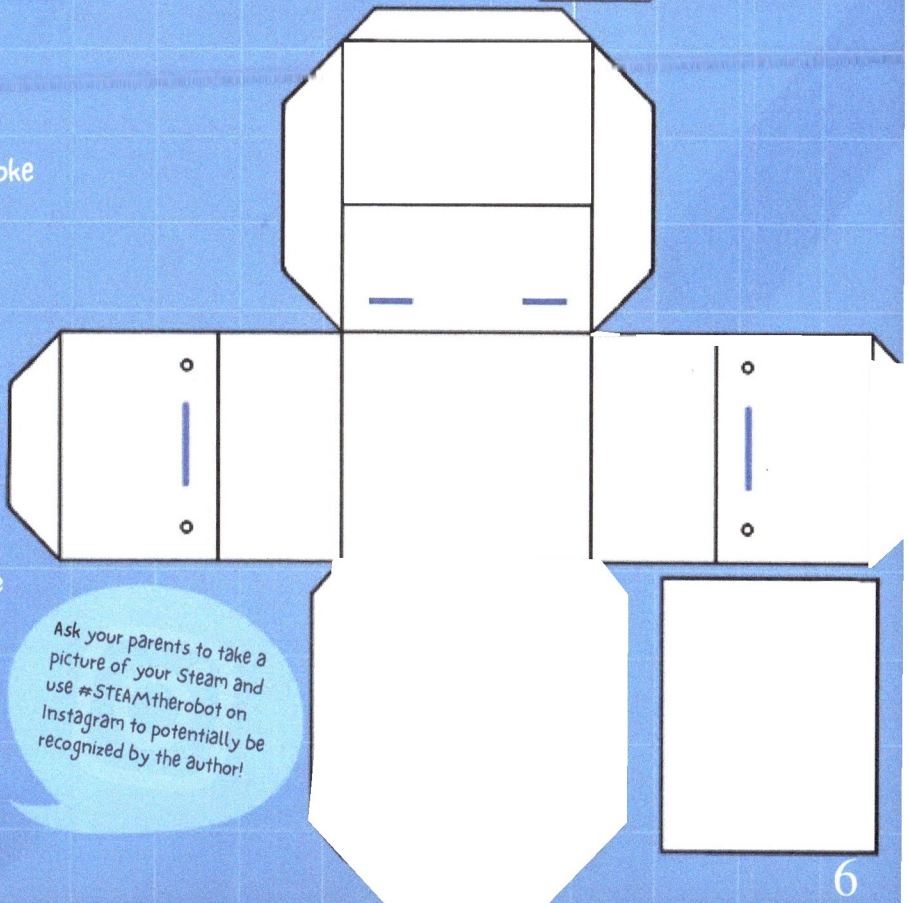

Ask your parents to take a picture of your Steam and use #STEAMtherobot on Instagram to potentially be recognized by the author!

6

After Amora created Steam, they began adventuring to different places in order to collect more tools for her creations. Although, the seasons caught up with them, the winter season began.

"Steam, look at all of these shapes on the windows!" said Amora as she moved closer to get a better view.

"those are ice flowers! They usually form on windows when exposed to very cold and very warm air," responded Steam.

"'Ice flowers?" that name sounds pretty strange," Amora said.

As Amora looked closer at the ice flowers, she realized each one differed from the other. She was intrigued by the variety, but knew that such a beautiful sight deserved a great name like Ice Flowers.

"I wonder what other cool sights there are to see..." said Amora.

Want to see some cool sights just like Amora did!? Now is your chance. Let's create...

CRYSTALS

Step 1: Pour very hot water and borax into a big cup (optional: add food coloring). then mix until, the borax is dissolved.

Step 2: tie one end of string around a popsicle stick. Leaving 3 inches hanging.

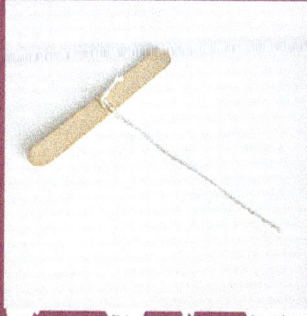

Step 3: Create any shape out of pipe cleaners.

Materials:
- Big cup
- Very hot water
- Borax (2cups)
- Popsicle stick
- Pipe Cleaner (1 or more)
- String (5inch long)

Optional: Food coloring

Step 4: tie the end of the string around your shape.

Step 5: Place the popsicle stick on top of cup allowing your design to enter the hot water.

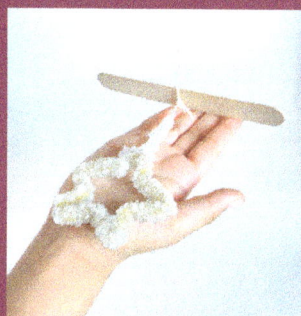

Step 6: Let it sit overnight and check out your design in the morning! (You can use it as a backpack keychain!)

10

After seeing something new, Amora wondered about other cool things she had never seen. At that moment, Amora knew she wanted to go on a quest; a quest to see new things.

"Steam! What if we traveled the world?" Amora asked happily.

"that's a very bold, but a great idea, Amora! I have just what we need, right here," said Steam as he pointed to his back.

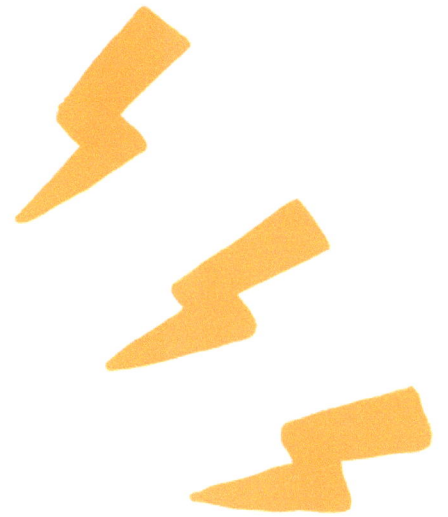

12

Steam stretched his arms to open a door on his back. Surprisingly, he pulled out the keyboard Amora had packed at the start. Steam explained that while they were traveling, he had slowly been modifying the keyboard and train! Now, the train could take them anywhere in the world.

"Where would you like to go? Any place we type in on this keyboard, the train will take us there!" Steam said.

keyboard! Were you expecting that? Well, guess what? Now it's time to learn some tips so you can type fast. This is important, as it means you can go on adventures more quickly!

Typing hacks: When typing...
Your thumbs should always hover over the "space" key.
Your hand should be in the same position as shown in the photo.
Keep your hands still, so your fingers can type faster.
Avoid typing with one finger! By practicing, you can type faster with all your fingers.

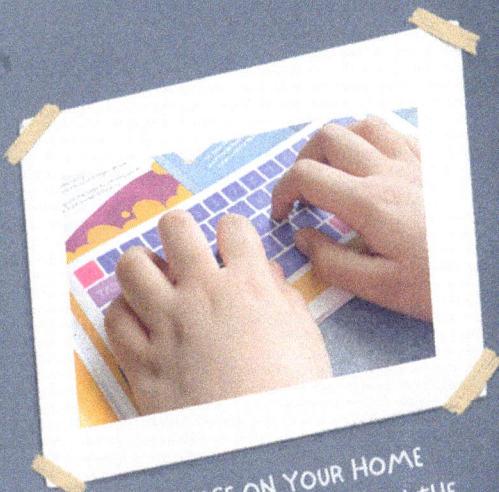

TRY THESE ON YOUR HOME COMPUTER! OR PRACTICE ON THE PAPER COMPUTER BELOW!

1 2 3 4 5 6 7 8 9 0 - += DELETE

TAB Q W E R T Y U I O P [{] } | \

CAPS LOCK A S D F G H J K L ; : " , RETURN

SHIFT Z X C V B N M < , > . / ? SHIFT

CTRL ATL SPACE CMD OPTION

Amora was excited. She quickly grabbed a globe and a magnifying glass, to find their first destination...

"Hmm...How about...Africa?"

Steam was open to anything! this is why without a second thought, Steam typed "Africa" into the keyboard, and off they went!

AFRICA

In the flash of an eye, the train arrived in a small town in Africa. The eagerness to see their first stop caused the adventurers to immediately start exploring. Eventually, they met a girl named tahani. She was welcoming and even showed Amora and Steam around her town. While showing them her town, tahani also discussed her favorite hobbies. One being, jewelry making; tahani even gave Amora a bracelet, to symbolize luck, wisdom, hope, well-being, and more.

Want to have a bracelet representing luck, wisdom, hope, well-being, and more?

Cut along the dark yellow line!

then glue the strip around your wrist like a bracelet! Now, you're done.

What do these patterns mean?

Prints are one of the most popular fabrics in Africa, worn by both boys and girls. Some individuals choose to wear Ankara fabrics in their Traditional African style, while others weave Ankara fabrics to match the western culture. "African Wax Prints," "Dutch Wax," and "Holland Wax" are all terms used to describe Ankara Prints. Ankara is a 100% cotton fabric with bright designs that are also known as "Ankara prints," "African prints," "African wax prints," "Holland wax," and "Dutch wax.". It's a highly adaptable fabric that can be used to make hats, jewelry, coats, bracelets, and shoes.

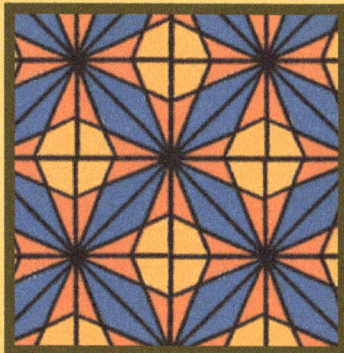

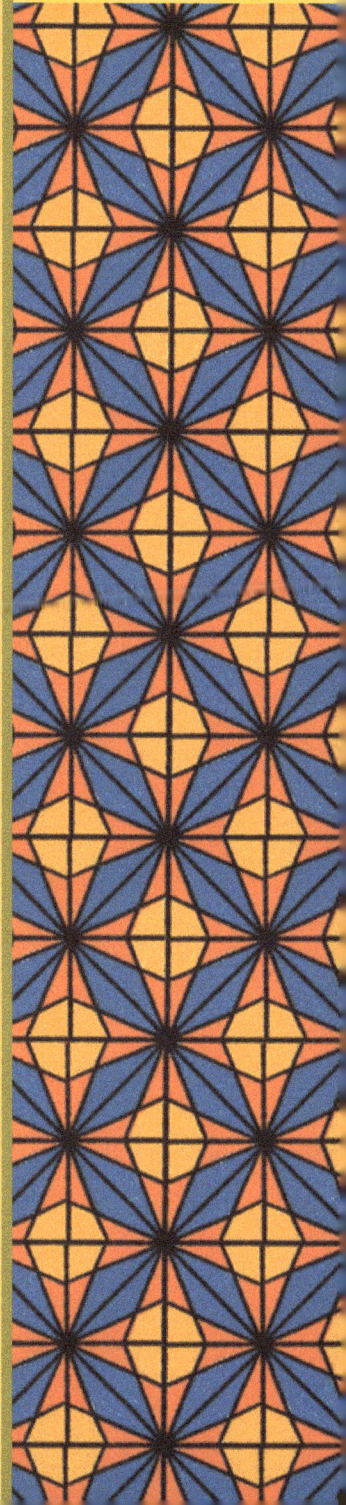

18

"Put glue on me!"

Amora proudly wore the bracelet as they finished exploring the town. Seeing all that Africa had made Amora happier than ever before. As tahani's tour continued, Amora became even more excited to see new places and people. So, she turned to Steam and asked...

"What should we type in next?"

Steam decided he wanted to venture somewhere south.
they went and looked at the globe to decide on a location.

"How about Australia?" said Steam abruptly.

Amora didn't wait to think, she went straight to the
keyboard and started typing in Sydney, Australia.

When Amora and Steam arrived in Sydney, a huge building was the first thing they saw.

"Oh wow, it's the Sydney Opera House!... One of the most photographed buildings in the world." Steam said to Amora.

"that's amazing! I wonder how they built such a huge and appealing building," she replied.

Strolling around the city, Amora and STEAM started discussing how things are built in such large dimensions. the idea of building functional things on such a large scale was fascinating to both of them. they wanted to know more! And after remembering how well their experience was with tahani, they went to find some help.

After walking around for a bit, they heard an abundance of noise coming from a building.

"I wonder who's making all that noise?" Amora said. Steam became curious as well, so they headed off into the building.

As they entered the mechanic shop, they saw a young woman who didn't seem super busy; So Steam and Amora approached her.

'Hello! We arrived today and want to explore the city. Are you willing to show us around?' Amora said excitedly.

RIV Motorsport

"Hey! I have lived in Australia for years, so I'm pretty confident I would be a good tour guide," said Rachel in her Australian accent.

"Sounds great!" said Steam.

When Steam spoke, Rachel looked surprised. Yet she didn't feel weird about Steam; there was something very friendly about him.

"Wow, you're such a cool robot! What's your name?" Rachel asked.

"My name is Steam, and this is my friend Amora. She built me!" he replied.

Rachel turned to Amora and looked at her with proud eyes. "I'm so glad to have met another engineer! It's hard to find those around here."

Rachel then checked her watch, and jumped after seeing the time "Crikey! It's getting late, I will show you guys around in the morning. Its time to close up shop for the night."

"Sounds like a plan! See you in the morning!" Amora said while she and Steam left.

24

Once they returned to the train, Amora and Steam decided to sleep on the roof; It provided a perfect view of the stars. They climbed to the top of the train with their sleeping bags and settled down. Amora started pointing out each star she knew, and Steam taught her about the ones she didn't. The young adventurers watched the stars as they awaited the morning.

After a good night's sleep, Amora and Steam met with Rachel so she could take them to the "coolest place in Sydney"...

"Welcome to the Coathanger!" she said as they approached the Sydney Harbour Bridge.

Steam was speechless. On the other hand, Amora ran around looking at every joint, nail, and screw that went into the creation of the bridge. She turned to Steam and said "Steam, we should build a smaller bridge kinda like this when we get back home!"

Your going to turn something big mini, By creating a small bridge out of unique materials!

Materials: Liquid Glue and LOTS of Jumbo and Regular popsicle sticks

Step 1-
Line up as many jumbo popsicle sticks as you would like until it is your desired length.

Step 2-
Glue regular sized popsicle sticks on the connecting parts of the other popsicle sticks.

Step 3-
Once dry, flip the popsicle stick structure over. then glue regular popsicle sticks on the edge of the popsicle stick row until it's covered.

Step 4-
Take two regular popsicle sticks and glue the ends together in a triangle pattern. REPEAT UNTIL YOU HAVE FOUR.

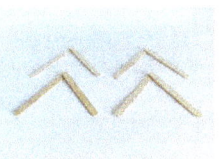

Step 5-
At the point of the triangle shaped popsicle sticks, glue a jumbo popsicle stick between them. REPEAT ONE MORE TIME.

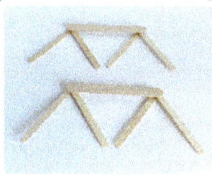

Step 6-
Glue the end of each triangle structure onto the sides of your platform

Step 7-
Glue four regular popsicle sticks onto the bottom of the structure and two at the bottom. SEE PHOTO ABOVE.

Step 8-
You're done! test out how much weight your bridge can carry!

28

Eventually, Amora, Rachel, and Steam went back to the train. Once they arrived, Amora thought about where to go next. During her **train** of thought, Rachel came up with a great idea...

"You know, if y'all liked this, you would love Thailand, also known as the 'Land of Smiles' there's great scenery and amazing people there." Amora and Steam looked at each other and smiled with eagerness. At that moment they knew their next stop was...

Thailand!

After arriving at the train, Amora took Steam's hand and led him onto the train so their journey to the "Land of Smiles" could begin.

When they arrived in Thailand, the adventurers were speechless; there were so many lanterns and flowers.

"Look Steam! All of these beautiful lanterns!" Amora said.

There was one store with many unique lanterns, too many to count. They were even selling the flower-shaped baskets that people were putting in the lake.

Amora quickly went into the shop. Once they entered, they were greeted by a young girl.

"Hi, I'm Rune! My dad owns the store but I can help with whatever you need"

"Hey, Rune! We were wondering about those flower things everyone keeps putting in the river," said Amora.

"Oh, the flower creations are Krathong's! We float them on the water during our annual festival; Loi Krathong. Loi Krathong celebrates letting go of worries, giving thanks for the things you have, and more."

"Oh wow, that's amazing! How can we get one?" asked Amora.

"Sure, I can show you how to make one!" replied Rune.

Rune began making the Krathong in front of Steam and Amora. They watched as she carefully folded each leaf. One time was enough for Amora; the whole process had been memorized in her tiny, yet mighty mind. She the started the process of making her Krathong.

Materials: Scissors, stapler, glue stick, pencil, ruler, colored paper (green is essential), plastic plate, and a fake candle.

Lets build a...
KRATHONG

1.

Use a paper plate to trace on green paper, then cut a circle. Now, add spikes to your circle.

2.

take any color paper of your choosing, and cut 2:6 inch rectangles (these will be your petals! So the more rectangles, the more flowers you will have!)

3.

Fold the rectangles to make a triangle.

Fold the paper to make a petal form. Staple it to keep it closed. (Repeat as needed)

4.

5.

Glue petals to green circle in a circle.

6.

Repeat to fill gaps. And continue until a gap the size of your candle is left in the middle.

7.

Glue your fake candle in the center.

Now watch your krathong float!

8.

36

With their completed krathongs, Amora and Steam sat beside the lake, enjoying a cool breeze. With Rune by their side, they slowly placed their krathongs into the water. As the krathong floated away, Rune asked, "Where are you both headed to now?"

Amora and Steam looked at each other, they wanted to visit one last place before returning home. Amora remembered the old Greek myths her mother used to tell her before bedtime.

"How about Greece?" asked Amora to Steam.

"Lets do it!" Steam said, and off they went.

they arrived in Greece just as the sun was rising. Amora and Steam took a moment to take in the colors of the sunrise. the peachy orange and bright yellow of the sunlight fell on the sea. they started walking around the busy streets of Greece, looking for someone who could show them around. Easily, they found a cheerful young boy named Lucas. well, it was more like Lucas found them.

"Hey, travelers! You look a bit lost. Need any help finding your way around town?"

"that would be wonderful!" said Amora. With that, they began walking along the streets of Greece with their new guide.

As Lucas was telling them about the history of Greece, one thing caught most of Amora's attention; the role of math in Greek history. Amora always talks about how much math wasn't her favorite subject in school but found it interesting how a group of people could create something so revolutionary. Noticing Amora's interest, Lucas told them about a famous Greek mathematician, Euclide.

THAT

Amora and Steam were quite shocked and impressed! However, he wasn't the only Greek to have worked with math. Lucas told them about more people; Like Archimedes and Ptolemy, who contributed to its discovery. After spending the entire night talking more about Euclid's impact on math, Amora and Steam went back to the train, now knowing lots of information.

Have you ever wondered how to make a perfect triangle? Well, here's a quick way to utilize Euclid's first proposition. Plus you'll learn something to utilize in the future.

Steps:

1) On this page you have two orange dots! Draw a straight line to connect them.

2) Now draw two circles, where the lines intersect. (You may see a faint guidance circle to help you!!)

3) Draw a line from dot "a" to the point at the top where the circle connects. Same with dot "b".

Now, you've made a perfect triangle! Try it out on a separate piece of paper without assistance lines.

a   b

EXAMPLE!

Amora woke up a little earlier than Steam, to catch the sunrise again. the beautiful colors in the sky gave her a relaxing setting to think in. She began reflecting on her amazing travels with Steam, and felt overjoyed. they had learned so much and seen so many new things...

However, it was time to head home.

Steam let out a big sigh, almost like he was saying goodbye to the sunrise. Sadly, he pulled out the keyboard from his back and typed "home".

On their journey back home, Amora looked out the window of the moving train. She was thinking about all the amazing people they had met and the very interesting places they had seen. She was glad her hard work had paid off.

As the train approached Amora's house, her mom and dad ran to the front porch. they were so proud of her journey and growth.

After meeting new people, exploring, and trying new things, Amora had become a young leader that was unlocking her true potential one step, one city, and one tool at a time. All while utilizing and learning science, technology, engineering, art, and math (S.T.E.A.M).

Amora's quest, or rather her S.T.E.A.M. quest, led her to learn new skills for her ever-changing future. All while taking her on a journey of discovery. Where can S.T.E.A.M take you?

THE END.

Final thoughts

thank you for reading STEAM Quest. I hope you enjoyed and gained something from this story! Exploring S.T.E.A.M (and with a little courage on her side) did great things for Amora. What can exploring new things do for you?

Always remember, to be courageous and try new things. Even if others aren't doing the same! You never know where it could take you.

ART IS A JOURNEY
OF DISCOVERY

active

is the beginning of

DO SOMETHING YOU

ATE EVERY DAY, JUST

FOR THE PRACTICE

Acknowledgements

"the best thing you could give someone is a chance"-Unknown

I want to thank...

My Nana: For listening to my rants about character names and dealing with my indecisiveness.

My Dad: For listening to my book updates and actively supporting my Instagram.

My Auntie Donna: For putting me in contact with my mentor, and always being supportive of me and my never-ending endeavors.

My Mentor: For being the person to keep me on track.

My Friends: For listening to my constant book updates.

My Family: For following my progress and for all the kind words!

thank you all again, for taking a chance, and assisting me throughout this journey.

About the Author

At the time of publishing, Ahmya is a teen passionate about unlocking the imaginations of young minds. In turn, assisting the journey of young leaders. She likes to describe herself as an advocate, mentor, and innovator. For the past 7 years Ahmya has been heavily involved in STE(A)M! During these years, she has created Nasa recognized research, built multiple robots, made coding curriculums, hosted international STE(A)M events, and much more. Through STEAM Quest, she hopes to provide accessible STEAM resources, educate youth on different cultures, and inspire students internationally to seek out opportunities for self-growth.

Fun facts:

-Ahmya wants to write 2 more books!
-Ahmya's a competitive chess player
-Ahmya is very excited to hopefully see your versions of Steam!!
(Don't forget to use the #, #STEAMtherobot on Instagram!)

Support

To scan, open the camera app on a phone and hold it up to the QR code.

Follow Ahmya on Instagram to keep up with her
and the future books she will make!

www.ingramcontent.com/pod-product-compliance
Lightning Source LLC
Chambersburg PA
CBHW040136200326
41458CB00025B/6280